Business Strategy for Water Challenges

From Risk to Opportunity

Stuart Orr
Head of Water Stewardship, WWF

Guy Pegram
Managing Director, Pegasys

Taylor & Francis Group

LONDON AND NEW YORK

First published 2014 by Greenleaf Publishing Limited

Published 2017 by Routledge
2 Park Square, Milton Park, Abingdon, Oxon OX14 4RN
711 Third Avenue, New York, NY 10017, USA

*Routledge is an imprint of the Taylor & Francis Group,
an informa business*

ISBN 978-1-910174-27-2 (pbk)

A catalogue record for this title is available from the British Library.

Page design and typesetting by Alison Rayner
Cover by Becky Chilcott

Abstract

WATER IS A RESOURCE UNDER INCREASED STRESS, with its management now cited as one of the greatest risks to business continuity and growth. It has come as somewhat of a surprise to many to see how quickly business and investors have started paying attention to this highly valuable resource. Yet there is still great confusion over how water is shared within society and with the environment and how its management is a complex and often under-resourced priority for government. With the advent of risk tools and a growing list of testaments around business risk from water, we are now able to plan and respond more appropriately to how this resource is used, impacted and impacts upon business. This book outlines these challenges and helps guide companies as they begin to build strategy around water.

About the Authors

STUART ORR's job as 'Head of Water Stewardship' is to work closely with business and public institutions to improve and prioritize the management of water resources. He leads a growing and exceptionally talented team inside WWF dedicated to proving that there are always better ways to develop and grow economies, protect the environment and work with business. He sits on a number of corporate sustainability boards, partnerships and think-tanks – all lovingly obsessed with the management of our most precious shared resource. Stuart has published mostly on water risk, measurement, agriculture, water policy and corporate governance. He lives in Switzerland with Jenny, Nerea and Luna.

GUY PEGRAM is a water planning engineer by training, but spends most of his time dealing with the regulation and management of natural resources and infrastructure. He leads a dynamic and multi-disciplinary management consultancy, Pegasys, through whichhe has spent the past 15 years addressing catchment water resources issues in Africa. More recently, he has been involved in global discussions around corporate water stewardship and the role of water in developing economies, which together pose the challenge and opportunity to build water and climate resilience into business and government activities.

Guy's home is in the beautiful city of Cape Town, which he shares with Greta and Luc.

....

Acknowledgments

THANKS TO the reviewers' helpful comments and suggestions as well as editing support from Jennifer Bew Orr.

Glossary

- **Freshwater** is naturally occurring water on the Earth's surface in ice sheets, ice caps, glaciers, icebergs, bogs, ponds, lakes, rivers and streams, and underground as groundwater in aquifers and underground streams. Freshwater is generally characterised by having low concentrations of dissolved salts and other total dissolved solids. The term specifically excludes seawater and brackish water.

- A **river basin** is the land area that is drained by a river and its tributaries. The Mississippi River basin, for example, is a vast area that covers much of the central United States from the central ranges of the Appalachian Mountains in the east to the eastern ranges of the Rocky Mountains in the west, funnelling toward its delta in southern Louisiana and emptying into the Gulf of Mexico.

- A **catchment** is the same as a river basin. Other terms that are used to describe a **catchment** are **catchment area**, **catchment basin**, **drainage area**, **river basin** and **water basin**. In North America, the term **watershed** is commonly used.

- A **licence to operate** is the need to gain and maintain the support of the people that live and work in the area of impact and influence of any given project – to have the social licence to operate. There is ample evidence that a failure to gain and maintain this social licence can lead to conflict, delays or cost for the proponents of a project.

- **Water governance** is the political, economic, administrative, social processes and institutions by which public authorities, communities and the private sector take decisions on how best to develop and manage water resources.

- The **water industry** provides drinking water and wastewater services (including sewage treatment) to residential, commercial and industrial sectors of the economy.

- **Water pricing** is a term that covers various processes to assign a price to water. These processes differ greatly under different circumstances and covers water prices for irrigation, utilities, urban use, etc.

- **Water scarcity** is the volumetric abundance, or lack thereof, of water supply. This is typically calculated as a ratio of human water consumption to available water supply in a given area. Water scarcity is a physical, objective reality that can be measured consistently across regions and over time.

- **Water stress** refers to the ability, or lack thereof, to meet human and ecological demand for water. Compared to scarcity, 'water stress' is a more inclusive and broader concept. It considers several physical aspects related to water resources, including water scarcity, but also water quality, environmental flows and the accessibility of water.

- **Water risk** refers to the probability of an entity experiencing a deleterious water-related event. Water risk is felt differently by every sector of society and the organisations within them and thus is defined and interpreted differently (even when they experience the same degree of water scarcity or water stress). That notwithstanding, many water-related conditions, such as water scarcity, pollution, poor governance, inadequate infrastructure, climate change and others, create risk for many different sectors and organisations simultaneously.

- **Enterprise risk** is a corporate governance process designed to identify potential events that may affect the business, and manage those risks according to the company's risk appetite, in order to provide reasonable assurance regarding the achievement of the company's objectives and strategy.

- **Water stewardship** (for business) is a progression of increased improvement of water use and a reduction in the water-related impacts of internal and value chain operations. More importantly, it is a commitment to the sustainable management of shared water resources in the public interest through collective action with other businesses, governments, NGOs and communities.

- The **water footprint** is the total volume of freshwater used to produce the goods and services consumed by the individual or community or produced by the business. Water use is measured in terms of water volumes consumed (evaporated or incorporated into a product) and/or polluted per unit of time.

- **Institutional investors** are organisations that pool large sums of money and invest in securities, property and other investment assets. Typical investors include banks, insurance companies, pension funds, hedge funds, investment advisors and mutual funds.

- **Socially responsible investing** (**SRI**), also known as sustainable, socially conscious, 'green' or ethical investing, is any investment strategy which seeks to consider both a financial return and social good.

- **Hot-spot** is a term given to a place or area that is thought to have higher levels of risks or issues that may need to be dealt with first.

- **Impact** refers to a higher level of pressure or negative change due to the use of water. Impact covers pollution and overuse of water or may refer to negative changes in biodiversity, access to water, etc. If water footprint is the amount of water used, impact tells more about the specifics of that use and in context.

Contents

CHAPTER 1

Introduction

IT IS AMAZING TO CONSIDER the speed with which water risk has percolated to top the business list of concern. Today, the World Economic Forum ranks the freshwater supply crisis as the number three risk affecting the global economy,[1] but already some companies are feeling the pressure closer to home. In April 2014, McDonald's launched their 2012–2013 Sustainability Report[2] to an internal McDonald's Energy Summit. Ken Koziol, Executive Vice President and Global Chief Restaurant Officer, took questions from the assembled audience. 'What is McDonald's water strategy?' was the first question.

David Grant, Sustainable Development Project Manager at SABMiller has a similar story. 'We were used to taking questions from SRI investors [Socially Responsible Investors], but we are now suddenly being asked by institutional investors, what are you doing about water?' After many years of companies telling us that no shareholder, investor or CEO has ever asked a question about water, things are changing. These remarks from SABMiller and McDonald's are just two examples of what numerous other companies are telling us.

Those companies already underway with strategies to manage water are finding themselves at an advantage. McDonald's, for example, estimated its corporate supply chain water footprint, looked at water risk for its restaurants, and used these findings in determining the company's sourcing priorities. McDonald's will soon be releasing their new water strategy. SABMiller has led on water for many years after having identified water as a long-term strategic risk. The message is clear: what companies are doing on water has become a top priority for their stakeholders and increasingly, their shareholders. It is certainly time for business leaders to consider water in ways that they haven't before.

Freshwater supplies and their management stand out as one of the more formidable challenges facing humanity over the coming decades.[3] In many parts of the world, water resources are stressed, and not only from a scarcity and pollution perspective. Fundamental issues relate to the governance and management of water, including investment, equity, ecological damage, loss of biodiversity and failed delivery of basic human services – the list goes on. But there are also plenty of examples where progress has been made and challenges have been successfully dealt with.

This book lays out how water challenges relate to business sustainability. It then proposes a strategic approach to water that will help you not only to understand this resource better but also to plan your own improvements, mitigate your risks and enhance the ways in which water is used and managed inside and outside your factory walls.

Some companies are already quite advanced in addressing water risks, while other are just getting started. What bears thinking through in depth are the ways that communities and companies share a complex 'root system' that feeds sustainable growth. If this book spurs a few more people to ask a new set of questions about water and to devise strategies and targets that will actually deal with the problems at hand, it will have met one of its fundamental goals.

1.1 **From risk to opportunity – from ad hoc to strategy**

To expand on what informs our perspective, it is worth explaining that the authors are career water professionals rather than sustainability experts. We have spent much of our working lives examining traditional aspects of water management: policy, accounting, regulations, governance, finance and planning, to name a few. We have come to Dō Shorts because the interest in and awareness of water issues that we see from business today is rising but is as yet largely unguided. To shift from traditional corporate sustainability thinking to *collective* water management thinking means recognising that the pressures, realities, risks and opportunities that business faces from water are mainly *external*. Therefore an understanding of how water is actually managed by governments and shared in society must supplement what has been done in improving internal efficiency. Companies need to know what has and has not worked before in water management. That is where our perspective has real value. Many businesses are starting from scratch in understanding water, when in fact much of the groundwork outside the often-referenced sustainability circles has already been done.

Water is a unique resource, constantly on the move through the global water cycle, as rainfall, rivers, streams, ice or evaporation. It is not something that we necessarily destroy by using it, but instead we use for a short while before it moves through the system to another use at another time. Globally there is enough water to meet our needs, but since water is the ultimate 'uncooperative' resource, this means we cannot ensure that it will arrive at the time and place and in the right condition we need it. Water is a local issue – and locally there are many aspects of water's management and use that are out of kilter with the natural cycle. So as part of our collective education in water risk and opportunity, it helps to establish water's fundamental difference from other natural resources.

1. Water availability is *variable* in time and location so its short- and long-term future availability is uncertain. One river basin may be suffering extended drought while neighbouring river basins are experiencing devastating floods. Equally, a given river basin may experience droughts and floods in quick succession. Understanding operational and strategic risk around water is therefore a different matter from understanding natural resources such as minerals or forests, which tend to be much more static.

2. Water is a *finite* albeit renewable resource, the availability of which is physically constrained by the infrastructure available, not to mention being legally regulated in many places by complex historical water rights systems. Differing pricing mechanisms and levels of governance of water systems all add to the complexity of managing this resource.

3. Water is *non-substitutable* in most domestic and productive activities. The risks associated with scarcity at the scale of a river basin are therefore very real. Put simply, while there may be substitutes for carbon in energy production, only water can be used for drinking or for irrigating crops.

4. Water is essentially a *local* resource. It is bulky and costly to move in the volumes typically required for production, so it can only be transferred between neighbouring river basins up to about 500 km (or even shorter distances where it has to be pumped

uphill). Because of these constraints, and because water flows from upstream to downstream users, risks and responses must be understood at a river basin scale – *not* at a global scale as is the case with carbon.

5. Water is *fundamental* to life and human dignity. Consider that rivers are part of a fragile ecosystem to which human settlements have historically been closely linked for transport, water use and waste disposal, and through their spiritual beliefs. Yet, these ecological, social and cultural dimensions are juxtaposed with the economic value of water related to its use in production processes. So more than with most natural resources, water management requires getting to grips with the risks around scarcity of an economic good, balanced with political risks given water's value as a social and ecological good.

This means that the insular thinking of yesterday's companies, framed around efficiency and technology, can no longer by itself prepare business for the future. Applying the same theory and response as has been used with carbon won't do it either.

It's fair to say that some companies have grown so large that their aggregate impacts could be compared with those of nation-states. The 58 companies that were analysed by CDP (formally Carbon Disclosure Project) in their 2011 Global Water Report alone represent a market value of US$2.49 trillion, equivalent to the GDP of a G5 country. These companies collectively abstract more than 1500 billion litres of water per annum, equal to 0.6 litres per day for every person on the planet.[4] That's just 58 companies. While most companies already have no idea of how large or small their current dependence on water is, to our knowledge not a single company has estimated the water they will need for future growth. How will you demonstrate your understanding of evolving dependency and risk around a shared water resource? How can you anticipate the changes (social, climatic, political, investment, governance, demographics) in those growth areas over time? If your investors aren't asking for that information yet, they will.

Water is part of the risk profile now and it's here to stay. In CDP's 2013 Global Water Report, 70% of companies report exposure to one or more water-related risks that could substantively affect their business. Approximately two-thirds of risks expected to impact on both direct operations (65%) and supply chains (62%) will materialise now or within the next five years. Yet only 6% of companies have targets or goals for community engagement, 4% for supply chain, 3% for watershed management and 1% for transparency, and no respondents set concrete targets or goals around public policy. That's two-thirds of risks approaching with NO planned response!

The supporting investors of CDP are taking note. Piet Klop is Senior Advisor Responsible Investment at PGGM Investments. He argues that for mainstream investors water is all about risk – that is, risk to their investment portfolio. 'Acknowledging that water scarcity and pollution can in fact put investment value at risk is a key step to take for companies, but especially for investors', says Klop. He concurs that not all risk can easily be monetised, but risk does impact on long-term investment value regardless. The risk to a company's continuity of operations and supply, risk to their future growth and risk to brand value must be considered in any reasonable profile.

What Piet Klop and other investors tell us they are looking for is 'a strategy that doesn't stop at improving operational efficiencies, but gradually evolves into efforts that increase companies' water security. The rise of external risk and insecurity demands collective action with other water users and authorities. Such a strategy signals that company management understands the systemic nature of water risk, as well as its role and possibly the business opportunities in mitigating such risk.'

So if this increasing awareness and evidence of water-related risk is coming through disclosure and from investor demands, then why have water issues maintained such low priority if not total invisibility within most companies? Some are acting on the issue, but even they have struggled to direct management attention towards achieving coherence on company strategy. Felix Ockborn, Environmental Sustainability Coordinator at H&M, had these discussions internally. 'Water is to many still not considered a *cost* and it is therefore not high on the agenda in many countries, where there often are more direct and pressing concerns, such as social issues. How would we turn this indirect and rather more silent risk into a serious priority?'

Very few companies will be able to 'sit out' the coming decade's water challenges. Whether located in the US or Bangladesh, whether selling running shoes or semi-conductors, almost all business sectors depend on water. More than that, almost all suppliers do too. These connections in our ever-connected world make your suppliers' risk your risk too. Thriving in an age of water constraints means addressing shared concerns about water, not 'fighting your corner' to the detriment of others. Even the process of creating a water strategy will lead to new questions and insights. Being at the beginning of something need not be so daunting if you understand that there are tools and guidance at hand. Others are also going through similar experiences and this is a sphere where cooperation rather than competition is the key.

1.2 **What is different about water risk?**

When we started out trying to incentivise better stewardship of water, it became clear to us that we were fundamentally missing the business case. By working with companies like SABMiller, H&M, Lafarge, Nestlé and The Coca-Cola Company, we learned a lot about how companies think about and integrate the issues. It was by learning how they communicate internally, how they needed to 'sell' water within their company, that we in turn became better at knowing how to speak to what matters. We also came to better understand corporate social responsibility (CSR) initiatives through working at first hand with companies in developing countries. These initiatives never quite added up for us. They weren't designed to address risk issues other than the obvious 'community engagement' category. At the corporate scale, water targets and goals often seemed to miss the mark. Either they were drowned out by huge sustainability pledges or they were reactive to what companies thought people wanted to see.

Ben Tuxworth, Director at Anthesis Group, a new global sustainability consultancy, agrees. 'Big name corporate sustainability strategies have become a hostage to fortune for some of the early adopters. It may be that the era of grand claims and big targets – and making a virtue out of not yet knowing how to reach them – is coming to an end. They are greeted with growing scepticism by stakeholders inside and outside the business, particularly where the link with commercial strategy is weak. For people who understand the real technical and operational challenges facing companies as they pursue sustainability in their value chains, the limits of unilateral strategies are pretty clear.' Tuxworth, like many sustainability leaders, recognises that water is a resource challenge that requires a more pre-competitive and collaborative strategy because of the systemic nature of water problems.

For the health of your business it's absolutely worth thinking through not just the attitude towards water and the vocabulary around it, but also the nature of the risk. A general problem is that risk depends on people's perceptions. As an example, a study[5] compared the deaths caused by kitchen germs and BSE (bovine spongiform encephalopathy – mad-cow disease to you and me), and domestic swimming pools and domestic weapons respectively. They noted that the 'risks that scare people and the risks that kill people are very different' (kitchen germs and swimming pools being the less scary but more dangerous in their respective cases). The study's authors note that risk could as easily be formulated as hazard + outrage, adding that adding that effective risk *communication* demands increasing outrage or urgency in accordance with the severity of the issue.

If we apply the formula of hazard + outrage here, there is very little sense of either of them concerning water scarcity or pollution. Water scarcity or pollution events tend to be un-dramatic and silent. If it's raining outside, it can't be too bad, right? The threat of water scarcity for example is often perceived as being in the future or being managed by 'those who control such things', unlike a terrorist attack, which is largely uncontrollable and unpredictable not to mention sudden. Accordingly, we associate terrorist attacks with high levels of outrage and perceive them as being a 'greater risk'. When hazard is high but outrage is low, people tend to under-react. More effective communication of how water impacts societies, economies, energy, people, businesses and the environment will be part of strategies that are geared towards action.

In 2008 we wrote a report entitled *Investigating Shared Risk in Water*.[6] At that time we were using the term 'shared risk' to encourage companies to look beyond their factory walls to the watersheds in which they the factories were situated. Shared risk was our attempt to prepare the groundwork for new ideas around water stewardship, and to start challenging the role of companies in public policy debates. We felt then, as we do now, that if companies could better understand water risks more from a stakeholder and river basin perspective, we could more effectively advocate for better water governance. Get the governance right, and risk overall is reduced.

But shared risk does not imply that we share *equal* risk or even the same risk. Risk from the subjective viewpoint should ask 'risk *to whom* and *of what*'. For the wheat farmer, the danger may be consecutive years of below average rainfall. For the manager of an industrial plant the risk might be a shutdown of water during peak operation time. For a government, risks might include the increasing costs of accessing water for utilities and the implications of higher energy costs, or failing to deliver on economic growth and development pathways because of poor water management.[7] Posit water challenges as shared and we begin to think about how to recruit interest, bringing those with common interests to the table. Shared vulnerability combined with effective communication can lead to a pooling of resources to solve problems.

At their most basic, these distinctive aspects of water mean that different risks and different facets of its use need to be addressed at different scales:

- Local level: a lack of drinking water, infrastructure, or competition over water.

- River basin level: allocation of water, infrastructure and management of floods and water quality.

- National level: policy, infrastructure and institutional capacity to manage water resources.

- Regional level: geopolitical disputes over water and energy.

- Global level: trade implications related to products and commodities with embedded water.

A key first message for companies is to understand that water risks are experienced first and predominantly by people and ecosystems at the local or river basin scale. Any successful risk management approach must be based on finding solutions that work not only for their business, but for local water users too. However, basic as this may sound, it does suggest a fundamental shift in thinking. Rather than being opportunistic, it is simply good business sense.

1.3 What is the state of our water resources?

The estimates of future freshwater availability, scarcity, pollution, flooding and management do not make for light entertainment, but it's essential to take on board the seriousness of the upcoming scenario in order to elicit the necessary urgency. The amount of freshwater on our planet is less than 3% of total water and less than 1% of this is readily usable by humans. Freshwater isn't nearly as plentiful as most of us think.

The 20th century saw huge advances in technology and in humans' ability to harness nature for productive purposes. In terms of water, societies developed infrastructure projects, for instance building large dams, to support irrigation, hydropower and industrial and urban development. This success in harnessing water had huge benefits for society but has come at a cost, with rivers and aquifers in many parts of the world drying up.

All manufacturing processes use water, another point that isn't common knowledge. And all products may be *viewed* as containing the quantity of water used in their production – referred to as 'embedded water'. At the same time, water is crucial to human life and for the survival of almost all ecosystems. The fact that water resources vary from place to place is also distinctive. We know that the world's water supply will become more stressed. Why? The short answer is change at a number of levels.

1. The world's population is expected to reach at least nine billion by 2050. The requirements for extra food and water are clear. Most of the three billion additional people will live in the developing world, where water resources are already stressed. Additionally, people are increasingly migrating towards cities that are already poorly served by water and sanitation infrastructure.

2. Rises in the standard of living in developing countries tend to result in higher per capita water requirements, especially through demand for different food crops.

3. Economic growth requires not only a direct water supply (for increasing domestic, agricultural and industrial use), but also a water supply for use in energy production (for instance, cooling water for thermal power plants, or river flows for hydropower).

4. Developing water infrastructure to meet water and energy needs results in the alteration of freshwater systems, creating potential conflicts between upstream communities and their downstream 'rivals' for water.

5. Climate change heightens water's variable nature through shifts in rainfall patterns and the melting of pack ice, often leading to reduced water availability in the current food and fibre producing regions of the world, as well as to increased flood risk to cities.

6. Collapsing wetland, river, lake and estuary ecosystems reduces resilience and the ability to directly provide flood attenuation, waste assimilation and food production.

7. An increasingly carbon-limited world could restrict the adoption of carbon-intensive technology in the longer term to solve water scarcity, such as energy-hungry desalinisation schemes and the pumping of water between river basins.

These changes are likely to be exacerbated by the lack of strong and politically independent water management institutions across much of the developing world, restricting our ability to use water effectively in such a changing environment.

One only needs to look at the condition of water resources in China to have a sense of the challenge facing growing economies. Water scarcity on the Yellow River, risk of flooding of large industrial areas in the Yangtze and severe deterioration of water quality as a result of industrial effluent, with consequences for health and treatment, have forced the government to prioritise the management of water, just as Europe and North America had to half a century ago. Yet even in places like Texas and California, water scarcity issues are making front page news. The lesson of course is that you never stop managing water. Adjustments must always be made over time as demand changes, with technologies, innovations and policies adapted to best allocate and protect water sources.

Freshwater scarcity and quality raises many fundamental questions for business too. Strong evidence suggests that companies cannot fully be insulated against societal water challenges. But in order to become part of the solution rather than part of the problem, companies will have to take a good look at their dependencies and grapple with some stark choices. Likely responses can be divided into three distinct paths that vary greatly in terms of outcomes. They are:

- Compete over water, risking conflict and protectionism.

- Commercialise water 'embedded' in agricultural products, which could result in increasingly variable and risky food markets, driven by climate change.

- Cooperate with others, aiming for increasingly equitable, efficient and sustainable management of water resources.

What can we do in order to actively bring on the last scenario? For one thing, we're looking a resource where 60% of the total lies in rivers, lakes and aquifers that are shared by different countries. Companies tell us all the time that cooperation and not conflict over water will be essential as they move forward. This is the good news, but we're still a long way away from seeing companies really internalise essentials around water. Lack of water can actually constrain growth in some areas, sectors or commodities, and choosing the right pathway from the list above requires deep commitment, knowledge and foresight. In certain circumstances when negotiating with other users, it may result in your getting less than you had before. Addressing water risk by managing it means that even if you're not able to secure as much as you want, the pain will be easier to take when you can plan for alternatives.

1.4 **What are we trying to achieve?**

Water management is a never-ending process of juggling competing needs. It is informed by the capacity and willingness around governance, measurement, regulatory change, investment, and infrastructure, to name a few. Growing needs must be reconciled with fixed water endowments and inevitably some countries are better at managing this than others. In that context, companies are able to navigate these water challenges in accordance with how mature the regulatory environment is, what type of recourse they have, and how engaged they are in terms of the needs and values of others depending on the same resource.

While managing risk is an essential element of business strategy, a company's first requirement is to better immerse in the real world of water issues. In many ways companies must accept some blame for how they have considered and responded to water (high pollution, overdraught of water, low understanding of the issues, poor compliance), having often lobbied *against* its shared management (dismantling regulatory controls, skewing policy).

The mining sector is one sector often viewed as having a lot of negotiating power. It's generally perceived as getting what it needs to develop and operate. Over the last few years however we have witnessed mining companies addressing water impacts and challenges faster than almost any other sector. Ross Hamilton is Director, Environment and Climate Change at ICMM, the International Council for Mining and Metals based in London. His organisation represents 22 mining companies, some of them the largest in the world. Hamilton states it very clearly: 'Water conflicts have been a wake-up call for us. The water permit you have, that piece of paper that says you have a right to use water, means nothing if other users are not ok with how you operate.' He agrees that his members are not necessarily on the same level as far as response, but by pushing in the same direction, they can reduce a lot of the complexity around water and create a clearer, more coherent way forward for the industry.

In response, ICMM has developed its Water Strategy and Action Plan, launching in April 2014 its Water Stewardship Framework.[8] The Framework explicitly recognises that water connects mining operations to the surrounding landscape and communities and provides a standardised set of guidelines to support effective industry response. It identifies four strategic imperatives: to be transparent and accountable, to engage pro-actively and inclusively, to adopt a catchment-based approach and to deliver effective water resource management. Further, it explains how a holistic understanding of the environmental, social, cultural and economic value of water enables material water risks to be identified, setting the context for effective operational responses.

One of the ICMM's member companies is US-based Newmont Mining Corporation. They have operations and mineral production in six countries and have faced some pretty contentious water conflicts in their day. Nick Cotts, Group Executive Sustainability and External Relations at Newmont, states that the real shift for his company came when they started thinking about water differently. 'Historically we tended to view water as a liability and a cost to our business, almost like a nuisance. Now we are seeing water as an asset. It is a resource that is valued by many different stakeholders and we have found that by valuing both the asset and the stakeholder we can find creative and shared opportunities to utilise these assets collectively.' Newmont has recently created a global water strategy. Now all of its operations are evaluating operational-related impacts and assessing opportunities to become better stewards of water resources.

1.5 Appreciating water's real value

When we talk about water and the private sector lots of images come to mind: water utilities, dam construction and operation, water-using companies and sectors (all with their own nuances in terms of knowledge, expectations, footprint and risk). There are plenty of people who make their money out of water, either through pipes, infrastructure and desalination (to name a few) or by buying up water rights and buying glaciers for water supply (we're not kidding). The water 'sector' therefore is huge, diverse and decentralised. We can therefore only provide the basic outlines of a water strategy.

Companies vary a lot in their response to water – some are steaming ahead on the issue, while others are setting meaningless targets with little knowledge of what real challenges lay ahead. As an example, a company we spoke to wanted to publish their water target as a 20% reduction in water use by 2030. Our first question for them was why? Who are you catering to? A regulator? An NGO on your back? Perhaps if that NGO doesn't understand water very well, they may think reduction targets are a great idea, when in fact they may not be. Reducing water use may be an appropriate strategy in some cases, but it doesn't necessarily reduce your risk. In a poorly managed water system that is prone to drought it may actually increase your risk. Let's say you improve your water efficiency to best practice levels. Then the local government slaps a 20% blanket restriction on all industrial water users during a dry period. Now you have a problem. The only way to manage this challenge (other than perhaps spending an unnecessary fortune on a technology fix) is to work with the regulator as you are building your efficiency, while looking for external opportunities that might benefit others at the same time that it helps you. Stay relevant, so that you are considered in future regulatory interventions. We're not sure this was the response the company was hoping for but it was really the one thing they needed to hear!

Right now a lot of companies think that because water is still cheap, that it is not a big problem for them. That's exactly what happens when you look at price and not *value* as your guide. When you only think in terms of how much water you need versus what your risks are. The new opportunities come if we reframe the issue, from 'this is causing me grief and higher reporting burden and cost' to 'it's in my best short and long-term business interest to see this resource managed well for society as a whole'. Reducing risk as a subset of greater opportunity really can bring a company forward in ways that haven't yet been totally accounted for.

....

CHAPTER 2

What Is Business Water Risk?

2.1 Regulatory, reputational and physical incidents and definitions

TODAY THERE ARE A NUMBER of risk tools to assess your water use and dependency. Now you can trace your use to river basins around the world, giving you a strong starting point for thinking about water challenges. At present the vast majority of companies still wouldn't know how to locate their primary facilities on a map, let along their suppliers or competitors. That's going to change.

....

FIGURE 1. Map of water risk.

SOURCE: WWF, 2014. View online at: http://www.dosustainability.com/wp-content/uploads/2014/10/water-risk-map.jpg

....

Felix Ockborn at H&M explains that once his company could overlay – in rough terms – their operations and main suppliers with areas of water scarcity on publically available maps, it opened their eyes to a new set of questions and risks. He says, 'The main question we asked ourselves was whether just continuing to work with improving water efficiency and ensuring waste water treatment in our supply chain would be enough. And exactly what risks are we actually subjected to? We've learned about reputational risks linked to practices in different parts of our value chain, but are we prepared for when operational risks such as lack of access to water resources or cost increases? How can we understand more about these regulatory risks that may come into play?' All are great questions of course and were generated by just looking at a simple map.

The Water Risk Filter (**http://waterriskfilter.org/**) and the Aqueduct tool (**http://aqueduct.wri.org/**) are two of the very best in helping companies get a sense of water risk issues. Of course, these tools are not just useful for individual companies but also portfolio managers, investors and donor banks. Piet Klop suggests that both tools '. . . are providing investors with more and more information on [companies' exposure] to decreasing water security. Index investors in particular hope that such information – through initiatives like CDP – will eventually make it to financial market information platforms such as Bloomberg, allowing us to compare and rank companies by their (aggregate) exposure to water risk.'

These risk tools cannot generate the 'perfect score'. They alone do not define what you face as challenges and how you have to respond. That can only be determined by more detailed analysis on the ground. But these tools do help start the process of water assessments for companies of any size by generating as much relevant information as possible and combining that with your location and behaviour.

While water challenges will provide business opportunities for some, it may threaten the operations or value chains of others. Although some are able to shift operations or supply chains with ease, others are

The water risk filter

The Water Risk Filter (**www.waterriskfilter.org**) was developed by World Wide Fund for Nature (**WWF**) in collaboration with German development bank Deutsche Entwicklungsgesellschaft (**DEG**). It is a free, online tool that allows investors and companies from all industry sectors to assess and quantify water-related risks. The Filter's risk assessment is based on a company's geographic location ('basin-related risks') and impact ('company-specific risk'). The Filter translates the most up-to-date scientific data sets, including the newest scarcity data, into risk metrics relevant for business. The results can be displayed on the company-wide/ portfolio level as well as on a facility level. In total, the risks are evaluated against almost 100 indicators. The tool also contains:

- Mitigation toolbox: bridging gap between risk indication and action on the ground

- Agri risk assessment for >120 agricultural commodities

- Country water fact sheets for all nations

- Development of industry specific indicators (hydropower, agriculture, pulp & paper etc)

- Mapping capability: ability to plot assessed facilities on maps with >500 different overlays constrained by political issues, by the need for market proximity, by the existing investment in a particular location, or by the location of crop production or minerals in specific river basins. All of these issues exist in a context of varying institutional capacity and political maturity of individual countries. The risk literature for business has also grown over the last five years, and has been helpful in terms of giving outlines which enable us to hone the narrative on water risk into three basic categories: regulatory, reputational and physical risks.

2.1.1 Regulatory risk

Many companies define regulatory risk as arising when regulations are too strict, or when compliance with them costs too much in company time and money. There is obvious truth to this scenario, but it's one that deserves closer examination. Surely regulatory risk would also arise from *not enough* regulation, or poor and inconsistent regulatory rules and norms? Again, most businesses thrive in a stable regulatory regime. Change, particularly when unpredictable, can be a serious problem. With water, we have to start thinking about how these aspects play off one another. Water regulations can be viewed not just as burdensome, but rather as presenting opportunities to mitigate potential risk. Water might actually be the one resource where business should crave consistent and strong regulations and even be willing to pay a little more money for that assurance. A full and integrated list that takes stewardship into account – meaning both the regulations for the company and those for the river basin – might look like Table 1.

TABLE 1. Regulatory risks

Risk – River basin specific	Risk – Company specific
Institutional weakness or management failure affecting quantity and quality of water supplies	Increasing competition with other users leading to water rights curtailment, disruption or revocation
Delayed government investment in infrastructure leading to restrictions during dry periods	Blanket industrial water use restrictions during drought that do not consider benchmarks
International basin at risk if other riparian state(s) have poor regulations and controls	Increasing costs for rights, storage, waste treatment and discharge
Local companies favoured over multi-national companies for licensing and fees	Government rejecting licences based on stakeholder concerns
Political concerns targeting companies' use of water through pressure on government	Inconsistent, unstable or unpredictable regulatory regimes
Poor or weak water pricing signals to reflect full financial costs, economic value and/ or social–environmental externalities in providing for sustainable water demand	Security of water rights, allocation or trading systems, ensuring that water for investment decisions will continue to be available

Companies often undertake efficiency measures internally without engaging externally with the regulator. We have spoken to a number of companies that have pushed efficiency measures to deliver record low water use only to have the external basin conditions affect their allocation. Simply put, they had built zero room to move. In catchments where there is plenty of water this may not be a problem, but in stressed environments, or in areas where water allocations are determined by political issues (we saw a company's intake halved – in Ireland!), it's crucial that the full range of potential regulatory risks are considered.

A great example of proactive regulatory engagement comes from Sasol, a globally integrated energy and chemicals company mainly based in South Africa. They recognised that due to water-stressed conditions, their water security was becoming a material challenge to operations in the Vaal River system. Studies by the Department of Water Affairs indicated that water shortages in the area could arise in the absence of significant action. Sasol uses about 4% of the catchment yield. Municipalities use approximately another 30%, losses from which can be as high as 45% due to the aging infrastructure. The company realised that by working beyond the factory fence, bigger advances could be achieved in enhancing water security in the catchment area, and at a lower financial cost. By investing in the municipality as opposed to their plant, Sasol obtained higher water saving rates, accrued the benefits they were seeking in water supply, and contributed to the wider community's water supply through improved municipal works – all at a fraction of the cost that would have been the case with internal technology implementations alone. Their anticipation of regulatory changes led them to explore water management practices external to their operations, with resulting gains in enhanced environmental and social impact and long-term financial returns.

2.1.2 Reputational risk

Reputational risk is the exposure of companies to censure and a resulting loss of customers due to perceived or real inequities around company water use or decisions. Reputation is one of business's most important assets, and also one of the most difficult to protect. It is harder to manage than other types of risk, largely because of a lack of established tools and techniques, and questions around accountability.

Risk can manifest when reduced water availability and quality give rise to tensions between businesses and local communities. Community opposition to industrial water use and perceived or real inequities in use can emerge quickly and affect businesses profoundly. Local conflicts can damage brand image or even result in the loss of the company's licence to operate.

The reputational risk to large water-using companies is greater where a water source in a river basin is in danger of habitat collapse,[9] or where water governance breaks down leading to a 'tragedy of the commons' type situation. Where this occurs and where scrutiny translates into public outrage, companies face dramatically amplified risks, especially when they are judged to be profligate or irresponsible. Where such crises unfold, there is a tendency for governments, NGOs and the media to apportion blame, sometimes fairly, but in other cases, somewhat opportunistically. High profile, multinational companies are usually easy targets for such blame regardless of their relative contribution to the problem, but local small and medium sized (SME) companies are not immune to this either.

Competition for water will increase, meaning more clashes will occur, especially where scarcity and pollution are outpacing action. Again, taking a basin and company approach, Table 2 gives some general observations on reputational risk.

TABLE 2. Reputational risks

Risk – River basin specific	Risk – Company specific
Poor performance and weak diligence in handling social and environmental concerns may lead to justified outrage from local or downstream users and NGO watchdogs	Concerns of stakeholders around quality and quantity from company operations can cause disruption or shut-down to operations or increase cost of doing business
Companies become 'scapegoats' for basin-wide water risk issues, even where they are not at fault	Depletion of the resource may create negative perceptions downstream in the same basin
Customers may choose not to source from a company due to actual or perceptions of how you handle water	Higher profile companies within the basin increase risk of being targeted if their interventions are not well executed

Certainly there are cases where the handling of water resources has been poor enough to backfire badly. Greenpeace released a report entitled 'Dirty Laundry' back in 2010[10] which looked at the problem of toxic water pollution by the textile industry in China. The report focused on two facilities that were found to be persistently discharging a range of hazardous chemicals. Greenpeace uncovered links between these local polluting facilities and a number of major international clothing, fashion and sportswear brands. High profile brands were accused of failed supply chain management concerning water quality. Most of the companies implicated are now actively seeking to reduce toxins from their supply chain. The companies should have identified and addressed these offences *before* Greenpeace placed them under the spotlight, but the moral of the story cannot be reduced to just 'don't get caught'. Some of these companies have hundreds if not thousands of suppliers, but even a basic hot-spot analysis and general water awareness might have told them where they needed to focus attention.

All companies along with other water users, contribute to the water situation in any given watershed. The water use of one entity and the overall water situation – the sustainability/health/resilience/status of the watershed – are far from synonymous. If local communities shut down your factory, will the closure of your plant solve the problem? Usually not. Local water problems are mainly systemic management issues, rather than single user problems, but then that's not the problem of reputational risk. The outcry alone can be damaging.

Where a single company is having an adverse affect on a watershed, reputations are justifiably tarnished. Companies and sectors which are heavy polluters and exceed localised limits or ignore stakeholder concerns should be taken to task and exposed. Equally, some who have taken an arrogant attitude to local values and stakeholder concerns have found themselves shut-out of decision-making. For those who dismantle public water regulations so they can have their own way, whatever you do outside of the factory gate will be from now on, under increased scrutiny.

2.1.3 Physical risk

Ultimately the above risks described derive from physical risk. That is, if there is low or no real physical risk (quantity or quality) issue, the others are unlikely to manifest. Physical risk is directly related to too little water (scarcity), too much water (flooding) or water that is unfit for use (pollution), all of which are associated with the management of a water resource. Such conditions may disrupt operations, reduce production capacity, or damage physical assets. Water pollution can impose costs on business by forcing facilities to invest for example in additional pre-treatment.

Risks can be associated with water resources at the river basin level, or at the supply level; namely sanitation and other infrastructure systems. Even where water is readily available, physical risk can emerge from poor management of the resource due to events outside the direct control of companies (Table 3).

TABLE 3. Physical risks

Risk – River basin specific	Risk – Company specific
Availability of freshwater is limited as a result of shifting demographics, economic activities or economic policy	Reliance on freshwater as an input and issues with water quality impacts. Assurance of supply jeopardised by failed governance and management outside of the company
Some basin users pollute water resources making water unusable or expensive to treat, affecting downstream users	Geographically fixed companies and supply chains cannot easily relocate
Climate change alters hydrology of basin and user needs	Disruptions of operations due to extreme weather events (floods or droughts)

Reliability of water supply and waste services to production sites due to poorly managed public water infrastructure operations or the capacity of public water managers	Flexibility in the supply chain, enabling sourcing of less thirsty substitutes or of products dependent upon less-stressed water resources

Agriculture is commonly seen as the sector most vulnerable to absolute water shortages. Farmers have highly risky futures due in part to their own exploitation of the water resource on which their businesses depend, but to an even greater extent due to the cumulative impacts of water use within a changing climate. Although on global average 70% of freshwater withdrawals are used in agriculture, industrial uses are also high. In California, for example, while the electronics manufacturing industry only uses a small proportion of state water use, it still requires a significant amount to produce even one silicon computer chip: around 8500 litres of de-ionised freshwater for its manufacture.

It's interesting to consider that physical water risk can also affect where you seek to sell your product and to grow your market. One consumer goods company we worked with displayed projected growth trajectories for one of their main markets: India. This growth was for products that required significant water amounts in the use-phase. The angles on the chart were like the side of the Matterhorn, almost straight up. On further questioning it became clear that the company's strategy division had never thought to ask about how water deficits or quality might impede the uptake of their product purchase and use: how their growth could be affected by a failing electricity and insufficient water supply. Questions we asked were, 'Is urban growth in line with water infrastructure delivery?' 'How much energy and water do Indian citizens for example, have each day to use for all their basic needs?' 'What is the impact of your product on water quality and pollution?' 'Could you be seen as part of the problem as opposed to a solution?' It became clear that a strategy for engagement in their largest potential market was needed. By being more visibly part of attempts to fix larger and more systemic problems, their largest market could yet be realised. Physical risks need to be understood not only from the point of view of operations, supply chain and customer use levels, but also within the water value chain as a whole, if your water strategy is to succeed optimally (Figure 2).

....

FIGURE 2. The supply and value chain aspects of water risk.

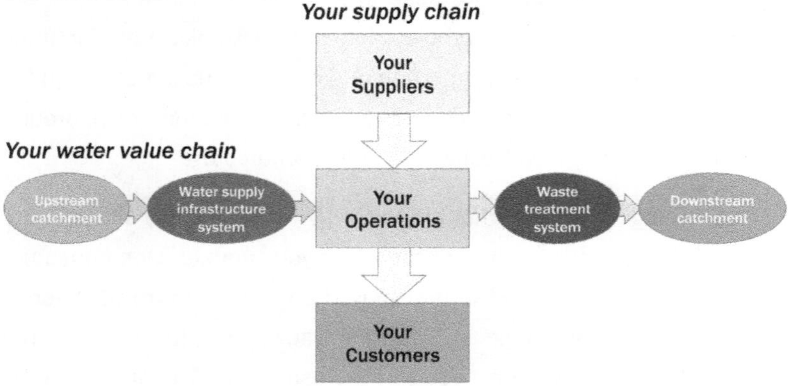

Your supply chain

Your Suppliers

Your water value chain

Upstream catchment → Water supply infrastructure system → **Your Operations** → Waste treatment system → Downstream catchment

Your Customers

....

Understand what this picture is telling you. Risk sits along your supply chain, but also across the *value* chain of water basins that you sit in.

2.2 Explore sector risk profiles

It's perhaps worth breaking risk categories into sector profiles as well. This helps in envisaging the commonalities within sectors because of their use and interaction with water. It also helps in assessing how the competition is responding. However, don't get obsessed by the competition. They may not be getting it right and the river basin should be a more important determinant of your strategic risk response than worries over what your competitor is doing. Figure 3 highlights the direct and indirect links to water which also affect risk exposure. Some broad water risk profiles are developed below for a number of business sectors.

- *Agri-business* is directly vulnerable to water stress and will increasingly face regulatory and financial risks. This is especially true where higher value uses of water begin to compete with agriculture and take precedence within a watershed. While many companies at this scale hedge these risks by shifting supply, there's ample reason to believe that this will not be the best long-term strategy. It also plays havoc with buyer pre-conditions and product-line pledges within the *food and beverage industry* which purchases from these larger companies. Their brand image creates reputational concerns in addition to regulatory issues incurred through the supply chain.

- *Consumer goods* industries are particularly vulnerable to reliability – quantity and timing – as well as quality of supply. They have potential regulatory and reputational risks due to water use and waste discharge resulting from both the production and consumption of their goods. *Other industry* and construction companies may be vulnerable to water availability in production.

Typically they have regulatory constraints around both water intake and waste discharge and their impacts on water resources linked with local communities. This has obvious associated reputational risk potential that can be leveraged through political means.

- **Extractive industries** experience regulatory vulnerability around the siting of mines. This is often seldom in areas where water is constrained (around rights issues or availability). Waste discharge is a serious issue with pollutant loads, flooding of sites, mine drainage issues and the storage of water. Mining companies have also been in heavy competition with communities in many parts of the world over supply and access – many examples of which continue to blight the industry's image. The *power generation sector* (particularly thermoelectric and nuclear) relies heavily on the availability of good quality water needed for cooling and at the right time, while hydropower relies on water flowing through their turbines. This sector has associated regulatory risks and is dependent on its privileged access to government as a strategic sector to maintain supply.

- The **water industry** is vulnerable to availability of supply, although acting for government typically immunises them from the physical risk. They've got significant reputational and regulatory risk around supply to communities and treatment of waste discharge.

- **Tourism and leisure** companies have to worry about health and availability of water in rivers and coastal aquifers, particularly in drier regions with dwindling supplies. Growing upstream needs and intermittent supply are also issues.

FIGURE 3. Direct and indirect water risk by sector.

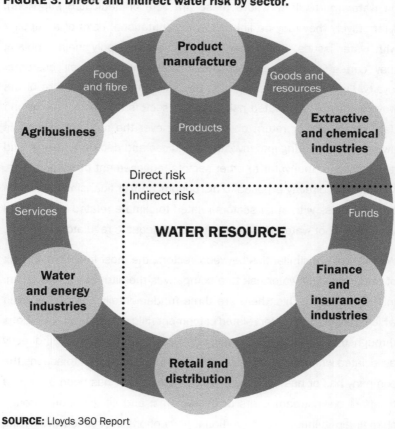

SOURCE: Lloyds 360 Report

In addition to the sectors above directly exposed to water risk, three other sectors have indirect exposure through their relationships with them. In many ways, it is these three sectors that have the influence and incentives that will make the catalyst for action.

Retail may be vulnerable to suppliers whose production or costs depend on water (particularly where these are sole or dedicated suppliers). Alternatively they may be linked to the reputational risks of a supplier with water issues. Even when the risks are a supply-chain problem, they can affect the retail chains' pricing and also their customer loyalty. *Finance and investment* is becoming increasingly aware of the way in which water-related risks in other sectors may pose risk around debt repayment or return on investment over the medium term. This awareness is filtering into investment indices and decision-making, and it a significant motivator to other sectors' engagement of water issues. *Insurance* is directly exposed to disruptions and/or financial impacts on their business with other sectors related to climate related hydrological variability, poor water management and inadequate regulatory regimes.

While there are similarities between sectors, the most important aspect of understanding water risk to a company is the situation in the basin. In thinking about this, there are three fundamentally different ways in which a site suffers water related catchment risk: i) directly on operations through the reliability and quality of supply, or flooding; ii) indirectly associated with regulatory or reputational impacts related to actions the company has or has not taken that has affected or has been perceived to affect downstream users or ecosystems; and iii) generally through changing regulations or reputational perceptions associated with being located in a catchment in which there are water related concerns, even where these are not related to your facility. Understanding and distinguishing these differences is important in developing targeted and effective strategy.

CHAPTER 3

Strategic Steps to Developing a Corporate Water Stewardship Strategy

3.1 The process of developing a corporate water strategy

MOST COMPANY STRATEGY is built around a cycle of planning, implementing, monitoring and reviewing. A water strategy shouldn't be any different. Find out first though how your company views risk and its appetite and approach to managing risk. Is it operational or strategic, or neither? Is there emphasis on proactive preventative, or reactive preventative, approaches to risk management, or both? Understanding the varied perceptions within the company on where risks lie, and how they are viewed, managed or mitigated helps in understanding how you can construct your arguments for support. As far as possible its development should be aligned or incorporated into the corporate or at least the sustainability strategy process in any company, as well as the corporate governance processes related to risk assessment and management. The following steps expand on the planning phase. They are similar to those used in implementing the accreditation process for the Alliance for Water Stewardship (AWS).[11]

Step 1 – Commit & Mobilise: Creating awareness and motivating the company's leadership around the development of the water strategy is an ongoing endeavour throughout the development and implementation of the strategy process.

Step 2 – Assess & Understand: Assessing risks and opportunities helps build the right platform for prioritising and planning. Learn which types of analysis are most suited to your company's water use in the entire supply chain.

Step 3 – Prioritise & Decide: As with all strategies, focus planning attention on the genuinely high risk issues. This may mean improving operational efficiency, addressing suppliers' use of water, or engaging local water management, depending on your sector and vulnerabilities. It is imperative that the water strategy has sign-off by the relevant managers and governance structures, so this is a phase requiring outreach and follow-up.

Step 4 – Implement & Communicate: Relevant targets of corporate and operational activities must be included when planning and budgeting. Personnel from internal communications and external marketing alike must be apprised of the ways in which your company is taking charge of its investments and disclosure.

Step 5 – Monitor & Evaluate: Adopting a water strategy means you have new types of information to support traditional operational and strategic metrics in a company. And just because you may not have risk metrics, doesn't mean you don't have risks. Bring water explicitly into corporate risk management and governance wherever possible, so that your company understands what is helping you move forward and where you are getting stuck.

Step 6 – Revise & Refine: Conditions and context may change rapidly in the water world, so keep adapting to ensure that plans stay relevant and effective, whether fine-tuning through implementation, or even revising the entire strategy starting from Step 1 after a couple of years.

FIGURE 4. The strategic planning cycle.

Water risks can affect a company at every level, though it may take quite some time before this becomes common knowledge. Once you begin implementing your strategy, there are certain questions that can never be asked often enough, namely:

- Are we doing what we said we would do?

- Is this achieving the outcomes we expected?

- If not, what have we learned?

- How is this impacting our exposure to water risks?

- Are there other areas that are posing new risks?

There are others considerations of course, but regularly re-evaluating, revisiting the objectives, truth checking and comparing against others, means that your water awareness and therefore your strategy will be fluid. You'll continue to take on board market needs and demands, especially from investors and shareholders. As water disclosure continues to gain traction, water strategies and response will be used to benchmark company performance.

Also on the rise, somewhat unfortunately, are plenty of opportunistic initiatives, labels and rewards for water initiatives that can only really be classed as gimmicks. This development was perhaps inevitable, given water's increasingly high profile. If you are being strategic there is less chance you will fall into a trap that values claims over strategy. This doesn't mean that you can't seek the occasional pat on the back or a dose of positive media though. But as stakeholders and shareholders get savvy about water, most of these schemes will be exposed for their weaknesses and failings. Save yourself the embarrassment, don't be tempted.

3.2 **Mainstreaming the water strategy**

As with all strategy processes, learning from *early* implementation improves understanding, confidence and possibly interest from management, informing the development of increasingly ambitious later-edition strategies. Managing water risk is a long-term project. Take your time. Understand the landscape. Begin with more achievable activities. Only transition to more sophisticated interventions with other stakeholders where it is necessary, and when you are ready.

Your corporate water strategy will be more effective if you can link it to broader corporate strategic imperatives. Your corporate culture around sustainability strategies will be a factor in the uptake of changes, but real conviction as to the profits of these labours is without a doubt the best motivating force you can have.

Are you primarily concerned with identifying water risk because your investors are worried, or because you have encountered risk first-hand? Is sustainability a values-driven initiative, or is it primarily for marketing purposes? In our experience, there are five inter-related reasons why companies engage with water risk and strategy. They are:

1. Reacting to an existing operational crisis

2. Envisioning future risks to operations or supply chains

3. Responding to external pressure from investors and consumer advocates

4. Recognising the competitive advantage in differentiating and marketing the company

5. Positioning in relationship to corporate social responsibility

It's very useful now to return to some of the earlier analogies of perceived risk versus the real likelihood of damage. The first two motivations cited above are based on incidents that have caught companies by surprise. These include media scrutiny, local protests, global brand campaigns, regulatory restrictions and even licence to operate. Perhaps if the companies were more attuned to external pressures and trends, they might have been able to handle the situations differently. However, it's equally possible that issues may have been out of their control. Perceptions or the failed management of public infrastructure are factors for which companies cannot take all the blame.

The third motivation, external 'market' pressure, has become a more recent factor as interest from the investor and disclosure communities rises. Also on the rise are actions by consumer advocates as they drive companies to become more water aware. The insurance and investment sector, for example, has everything to gain by provoking their clients towards a new set of actions around water risk and response.

The last two motivations have to do especially with promoting 'green' credentials. CSR initiatives must be strategic when it comes to risk reduction. Haphazard attempts at 'media hits' are worse than useless. Where legitimate crises exist and water risk is absolutely genuine, CSR must be coordinated as PR (public relations) that is founded on progress. That progress can only be realised by informed strategy. If not, a company can shift from a relatively secure profile to one that reflects a major water crisis – very rapidly. The degree to which the outcomes of the water strategy are linked to operational activities, as well as the uptake of water risk indicators in corporate risk monitoring systems will ultimately influence whether water risk is effectively mitigated.

3.3 **Strategic framework for water stewardship**

Much as real strategy trumps media hits any day, your actual framework of actions and priorities will help you hone in on legitimate priorities. It will keep you and your company aligned internally while moving towards the goals you set, proving that you're about more than greenwash. Figure 5 and the corresponding text is a framework to help mapping your strategy for engaging water risk. This has proven to be very useful in helping the companies we have worked with so far, and may assist you to think through the drivers and opportunities for managing water issues inside and outside the factory fence line.

FIGURE 5. Water strategy elements.

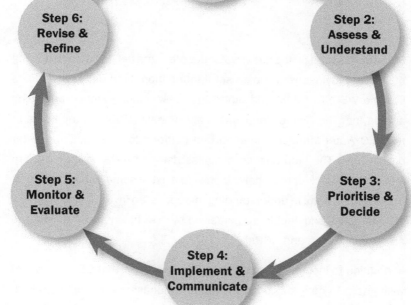

SOURCE: The authors

A framework should be seen as an evolving engagement with water stewardship. It should indicate the types of initiatives most appropriate to you. It is not necessary and non-advisable to attempt to tackle all five elements in your first year. Rather see these as the possible ingredients of the strategy. The role of each ingredient takes shape in accordance with your own learning curve. Here's what you will need to think about as you continue to educate yourself on water's relevance to your company.

3.3.1

Being up to speed on the latest water debates, and not just those affecting business, provides your most useful foundation. Find out what entities like the WBCSD, ICMM and other business/industry groups are doing and saying. The United Nations Global Compact (UNGC) is the home of the CEO Water Mandate, an important platform for business and water discussions. The literature on business risk and water is growing. Tap into it. At the back of this book we provide a list of some publications that we've found useful in understanding this space, so invest in knowing how your company and sector are perceived by river basin stakeholders, the press, consumers and NGOs.

In addition to increasing your knowledge, recognise that CEO and board willingness to optimally position relevant managers and suppliers is of great help. Having leadership on board means that companies can 'sell the water story' internally and better support any identified concerns. It's an axiom in water as well as in general that knowledge is power – but so are human resources!

Some key questions can help you know how you know how your operations and supply chains can put you in conflict with long-term success:

- Where does my water actually come from? Who is downstream/ upstream?

- Who are my neighbours? Where do they get water from? Is water a primary obstacle for them?

- What are the trends (climate, policy, demographics, etc.) in these river basins with high risk?

- Are policies, regulations, permits and charges likely to change?

- Are there catchments we work in that are becoming more stressed or polluted?

- Who's in charge of managing the water I am reliant on?

- Who's the target audience of my commitments on water?

- Which NGOs, law-makers and business groups are worth learning from? Which are not?

- How well has that strategy served us – do we need to do more? Or different?

- Are we measuring progress correctly?

3.3.2

Companies taking the first steps in looking at risk and water footprint issues begin by gaining a wider understanding of where their factories and suppliers are located (it's amazing how many don't know this and how easy it is to do). This may include a water footprint assessment, as well as some measurement of the impact that a company's activities have on water (plus the effects on people and ecosystems).

Water Footprints

The advent of the water footprint has allowed many companies to ask new and far-reaching questions about their water use and impacts. Water accounting and productivity analysis have long been used by water scientists, but work by Dutch academic Arjen Hoekstra to develop an accounting framework for the 'embedded' water in products has really helped place water onto the business agenda. In 2011, the Water Footprint Network (WFN) created by a collection of businesses and interested organisations published a manual that laid out for the first time, a standardised approach to water footprint accounting.

Water footprints are a useful way of assessing water dependency and vulnerability, and provide a basic measurement of impact along the supply chain. A compelling case study by SABMiller and WWF established that 98% of the company's water footprint in South Africa was related to crop production. However, there is a real challenge in getting too obsessed about the numbers rather than the underlying risk narrative. Again, unlike carbon, the objective of a water strategy is not to calculate the footprint and then drive down the numbers. Understanding water risk means using these metrics to explore relevant questions and highlight hotspots, rather than spend excessive time and precious resources refining the metrics. The search for the 'pefect' number – is a fool's errand.

The next step is to learn more about what this means in terms of the dependencies and challenges you may have – both in terms of quantity and quality. Map out and understand the wider context of your water use, including peer examples and more material risk issues. Explore the context of specific river basins, and the identification of high risk 'hot spots' caused by water quantity or quality issues relevant to the company.

Draw on the tools designed to make your understanding of water and business an application as well as a concept. Online water risk and footprint tools can help you to flag key areas and issues, as you build your arguments and business case. The databases and catalogues we've listed also serve as useful resources for learning what others have got right and wrong when it comes to reputational, physical and regulatory issues.

Swedish clothing retailer H&M's experience mirrors that of others we've spoken to in the sector. Felix Ockborn, Environmental Sustainability Coordinator, recounts some of the early conversations that H&M had around Step 2. 'In production, people asked if we should phase out all wet processing in water risk areas. But there are many different types of risk parameters, so which one should we use and where would we draw the line? And was this really the best and most responsible way to deal with the risk? We had been part of building up capacity for textile production in many areas which experience water risk but were there other ways of reducing water risks in these areas to enable the textile industry to continue being a positive driver for economic development?' These kinds of questions from amongst staff and champions within the company help pave the way for designing the right response.

In terms of what companies have got right, there are some useful examples in place. As a food and beverage company, SAB-Miller is particularly aware of the water required to grow the barley and hops for brewing. Now they've applied a risk model to understand the water used in the production of these crops as well as the vulnerabilities in the places they are grown.[12] Within the same sector, Coke introduced their Source Vulnerability Assessments to assess water supply and waste discharge vulnerabilities. In mining, ICMM is in the process of developing guidance for its member companies to assess the catchment impacts of mining activities. All of these industries are moving beyond internal indices alone with a view to securing legitimate positions in respective basins.

When it comes to guidelines for thinking through your unique case and the assessments you need to generate, there are the five main aspects you should cover (these are also the basis of the Water Risk Filter and emerging disclosure systems):

- How is water used in your production process and how does this compare to other similar operations (benchmarking)?

- Have you got your existing or planned water abstraction or waste discharge permits, and are any conditions (including charges/tariffs) likely to change in the near future?

- Is the water infrastructure required to assure your supply or treat your wastewater adequate? Is it being managed effectively?

- Are there any major water concerns related to local communities, other users or the environment in the catchment/s within which you operate?

- How is water user by your suppliers or customers, and are there any supply, discharge or catchment related issues you need to be concerned about?

3.3.3

Once you have a sense of where the risk issues are, the logical next step is to outline actions, targets, goals, and plans to tackle the more immediate solutions. It can also be a good time to promote wider awareness throughout the company. Internal action means engagement with employees, buyers and suppliers to establish the potential opportunities for the company. Water efficiency (where appropriate), pollution reduction, measuring and reporting, and internal water governance are all crucial elements of internal action. A pragmatic risk management approach also distinguishes preventative measures that can be put in place to avoid a risk, from preparedness plans that can be implemented if a risk occurs. The selection of relevant internal interventions or controls will depend upon company appetite and approach to risk, as well as the costs and benefits of different options given the likelihood or its occurring and the impacts if it does.

What internal interest in the subject can you identify? What resources are available to you? What can you do to respond to the priorities you have established through your risk assessment? Set your goals and objectives, identifying actions toward developing, publicising and communicating the strategy. Getting these internal actions incorporated within operational budgets is just as important as identifying the risks and responses in the strategy. Remember that you have to 'get your own house in order' before gaining the credibility to tackle other stewardship responses.

Of inestimable value is making sure that employees have the skills, knowledge and reward mechanisms to actually deliver what you need. There's nothing worse than seeing a plant manger being asked to engage with local communities around the factory. It's neither their skill nor their job. When setting your goals and targets it's important to ask, 'who are these targets for' and 'do we have the right skills mix to deliver'? If you need to disclose for investors, what type of information are they interested in? Or are the targets meant to reach business award committees? Make sure that what you are setting out to do is what people, shareholders, investors, communities, regulators and NGOs actually expect of you. Almost all of the companies currently leading comprehensive water stewardship engagement began with internal action. Your first couple of years will include a lot of it.

It was only a few years ago that companies articulated their targets solely in terms of water efficiency. This was *only* at a factory level: waste water reduced, water 'saved'. While some targets genuinely require technical expertise, many will be dictated by engineers fighting over who can run the best numbers. Your water strategy should, over time, help the company shift from an internal management focus to one of external water stewardship and either align or collide with societal and environmental concerns. In an increasingly water-stressed real world, the 'outside world' component we've described has far more impact than has yet been taken into account. It's where stakeholder expectations and perceptions begin to have traction in a strategy that acknowledges the realities of their existence.

The first three steps in Figure 5 are materially distinct from the next two. The first three steps have maintained an internal focus. The next two are where a company shifts from management to stewardship – where the rules, measures, focus, engagement, control and complexity change considerably and where traditional notions of business sustainability are most challenged by water resource realities. Too often we see companies bringing their internal strategies 'outside' without really knowing the realities of their challenges. Over time they come to see how inappropriate their actions are. Table 4 gives some thought to how the internal and external worlds should frame your thinking.

TABLE 4. Internal and external differences in the strategy.

Steps 1, 2 and 3	Steps 4 and 5
Direct sphere of control	Indirect sphere of influence
Impacts my company has on water resources	How my company is impacted by external water issues
Efficiency of resources	Allocation of resources
Products I make (or buy or use)	Places I (or others) make them
Private goods	Public goods
The value I create	The values people hold
The risk I face	The risk we face

SOURCE: WWF, 2013

3.3.4   ④ Collective action

In some cases, internal actions in operations and supply chains may not be adequate to reduce your company's risk to acceptable levels. Collective action reflects that a company works with others at various scales as part of its strategy. Today, companies are finding that technical fixes alone won't solve their long-term issues. Anheuser-Busch InBev focuses on 'reducing water risks and improving water management in risk areas'; H&M seeks to 'work together on collective stakeholder engagement and water management forums in prioritised river basins', and General Mills are 'implementing changes in high-risk watershed areas' and 'developing public commitments, public education and advocacy with watershed neighbours'. This shift in external commitments to more strategic approaches will bring company actions into more 'mainstream' water debates, creating pathways for others.

Collective action's emphasis – working with other stakeholders – enables strategic external engagement and negotiation that meets common interests. It can support others with limited resources. It widens the agenda to better achieve your ends. Stakeholders can be anyone from other users within a geographical area, such as a specific water catchment, to other companies, communities, sector initiatives, public agencies, NGOs and standard-setting bodies.

Collective action may sound vague, but the CEO Water Mandate has developed a Guide[13] that distinguishes levels of collective action. These range from actively sharing information through to joint decision-making and implementation. The focus of this action may be on improving infrastructure operations, strengthening regulatory decision-making or fostering catchment management. All of these elements directly affect company water supplies and waste treatment. Again, it is worth noting that collective action may be designed to prevent a risk from occurring in the catchment, or alternatively prepare to mitigate the eventuality of the risk at a local or catchment scale.

It's fair to say that showing attribution for collective action – being able to take some credit for progress – is difficult. It implies a need for support from leadership, who need to be made aware that guidelines for this exist and that the indices for this approach are being developed. In fact, even Anheuser-Busch InBev, H&M and General Mills admit that they're on a journey to work through the right indices. Yet knowing that their future risks can only be dealt with by being proactive has emboldened the frontrunners to work through this new approach and is helping others to see what can be achieved.

3.3.5

When thinking through who the players are in the process of governance, consider also what their motivations are. Water governance refers to the political, social, economic and administrative systems that are in place to develop and manage resources, and their delivery at various societal levels. Often when we refer to governance we've got the narrow perspective of national government in mind – its policies, laws, regulations and allocation of resources. But improving governance also happens through engagement with all sorts of stakeholders outside of governmental capacities. Water stewardship can be considered as actions by water users themselves to contribute to the management of the shared resource towards public good outcomes.[14]

As for why all these stakeholders feel the need to engage, there's a strong common platform: risk and uncertainty. If public authorities are seen to be amongst your risk factors, engage with them. Again, the CEO Water Mandate has developed a Guide to help you. Companies working with other companies on these initiatives serve credibility, and increase real world impact on governance.

Engagement activity depends not only on the sector but on that sector's ability to influence. This is the case whether the entity is a strategic partner of government (energy, water provision) or a manufacturer of goods. Once the engagement begins, its success requires that the business position be aligned with broader public interest. This type of engagement may involve lobbying independently or with other companies. Whether we're talking about risk mitigation in specific locations or as part of broad coalitions of businesses and NGOs, all of these efforts will help create greater political support for progressive water legislation and implementation.

3.4 Specific considerations for the corporate water strategy

One of the more confusing spaces is that where water and business debates collide: too many acronyms and initiatives that overlap and even seem to contradict one another. Tools, approaches, standards, footprints, etc. What's good, bad, ugly and what's useless?!

So it's helpful to be reminded of CDP's 2013 report which showed that companies say they know they have high risk, but aren't setting the right strategy. This is partly because of the confusing landscape, and partly because of lacking strategy. If companies were spending the time to map the water stewardship landscape and settle on where their risk issues were, then there'd be less confusion. Reflect on what others are doing, but don't let that necessarily dictate your every move. That's part of your Water Awareness. Consider the big distinction between an *internal* strategy, a *supply chain* strategy and an *external* strategy and understand the differences between them. Also, remember your water value chain (Figure 2). Be aware of the measures working within the factory walls, but also that once you get outside the fence-line of your operations, a new set of valuations become relevant.

Efficiency and process engineering – For some time now, companies have been driving efficiency within the factory walls and seeking technical devices and solutions to reduce water pollution, hopefully saving some energy and cash in the process. Yet as noted before, efficiency doesn't necessarily prevent restrictions from being applied to you. Nor does it entirely account for how stakeholders will view your handling of a resource that's vital to them too. An overly strong focus on driving efficiency can steal focus from what really matters. Risk more than footprint should be the ordering that drives strategy.

Disclosure – Companies disclose for a number of reasons. One is in response to market signals from investors and shareholders. Another is to attract funding. Still another is to be transparent. Disclosure can actually help to sell water within your company and potentially to benchmark against industry peers. It enables the sharing of best practice, increasing market share. It indicates how well you are trying to perform, and can work in your favour when future water licence applications or restrictions are applied.

Supply chains – For many sectors, most of your water footprint or risks reside right here. Some companies dive deep into the lower tiers of their supply chain for reporting, while others are still struggling to know where they source from. With online tools freely available today to map and request information from suppliers, not knowing your terrain can't be an excuse. For some companies with 10,000 suppliers – with each supplier using water in different ways and in different places, addressing the few suppliers with the greatest risks is crucial.

Use phase – For the makers of detergents, cleaning products, chemicals, washing machines, etc., developing technologies and approaches that reduce the amount of water used or waste produced, is key. In particular, pay attention to where markets in which your customers face real and almost daily struggles in getting reliable, safe and affordable water and energy.

Social licence to operate – Local engagement in areas where social and environmental problems pose risks to a business provides a testing ground for understanding social licence to operate. Company reputations depend on meeting targets around stakeholder engagement. Positive social development casts companies in a light in which you want to be seen. Yet often local engagement has been somewhat disconnected from strategy – more an ad hoc response to a crisis. That doesn't make a response wrong, but it does begin to tie a company into potentially reactive positions and into projects that they can be bogged down in for years. Again, good for the CSR report, but good for actual risk reduction? Better to be strategic.

Catchment governance – All operations and supply chains are located in water catchments, usually with a range of other water users and interested stakeholders. Decisions about how to manage and regulate catchments are increasingly being informed by diverse external perspectives. This means that the result of sustainable water management actions by government or public sector catchment authorities is typically the subject of wide consultation. Fostering inclusive and effective catchment governance can help you build consistently applied, coherent and predictable management and regulatory regimes and is therefore serves the long-term interest of a company's stewardship goals.

National and global leadership – Being part of the broad water policy debate at a national and global scale helps position a company as a market leader. It ensures that the enabling regulatory environment is in place and yet in and of itself, it does not reduce risk. It must be supported by local action in real places. This builds credibility around your company's leadership, and genuinely informs the larger position you seek. Link leadership aspirations with tangible results in the places you want to influence.

3.5 Some potential risk implications of external response and engagement

Be aware that engagement in policy dialogues has an associated risk when not coupled with transparency and inclusivity. Smaller companies or civil society groups may be sensitive to possible policy capture accusations and realities that are present where business engages directly with government to influence public policy. This is part of why collective action is so important: it's how you prepare the ground for the realities you influence.

So while there are compelling arguments for businesses to address their water risks by engaging on external water governance, there are potentially new risks from failing to approach governance in the correct way. The importance of water to the environment and communities, as well to issues of food and energy security, means that water policy and its implementation are ultimately a government mandate. The perception of policy capture invokes major reputational risk. Companies need to value transparency and wield their influence judiciously.

The CEO Water Mandate's Guide to Responsible Business Engagement with Water Policy sets out five principles that should inform any company action beyond the fence-line. They are: advance sustainable water management, respect public and private roles, strive for inclusiveness and partnerships, be pragmatic in considering integrated engagement, and be accountable and transparent. These are obviously broad considerations, but the Guide does a great job of talking through these in detail, explaining the do and don't scenarios around each principle.

As all risk managers know, intervention beyond the direct operations of a company may mitigate one set of risks, yet incur a new set. This is the case in the highly politicised world of water – the following areas should be considered in any strategy.

1. Interventions that are not aligned to local/river basin needs create a risk of being ignored or even opposed by stakeholders.

2. Absence of key political or stakeholder engagement can expose the initiative and the company to later accusations of 'railroading'.

3. Effective engagement requires skills and expertise from both companies and associated stakeholders. Lacking expertise in government or in stakeholders jeopardises the ability to engage.

4. Perceptions around corporate influence on government – either

lobbying solely for their own interests or to the disadvantage of other groups – can sometimes mean as much as the realities.

5. Government abdication of responsibility for water functions to business leaves the company with responsibility for a non-core function.

6. Human and financial capital requirements increase as engagement becomes more complex.

7. Making an exit (from a project or location) poses challenges once effective engagement has achieved the corporate objectives, particularly where this has led to an intervention (i.e. policy capture).

These risks fall into two broad categories, namely the legitimacy of the engagement (in the case of the first four) and the ongoing responsibilities associated with the engagement (the last three). The former can be dealt with by engaging in good faith from a shared perspective and entering into partnership with relevant NGOs. The latter requires comprehensive strategic planning before embarking on any process of engagement. These risks are particularly relevant in developing countries, but similar issues may also arise in managing risks within industrialised countries.

CHAPTER 4

Going Forward –
Opportunities

THE ATTENTION TO FRESHWATER and its management from the business world has come as surprise to many traditional water practitioners, but as we keep saying to anyone who listens, 'this isn't going away'. Questions about the future provoke challenges in three areas of water stewardship: the resource, the response and in staying on course. Increasingly progressive companies understand that acting jointly in places with water-related concerns represents the next significant step in the stewardship journey. Early initiatives from within different sectors reflect certain trends:

- Sectorally, mining group ICMM, for example, is advancing an ambitious strategy with their members. Over the coming year they will be putting together guidance for their members to individually and jointly assess and engage catchment management, potentially seeking to link with other mines, sectors, communities and government.

- WWF is assessing river basins around the world to prioritise those in which water and economy make a compelling case for convening basin action between business, civil society and government. This will enable large collective action approaches across sectors with local government and stakeholders, linked to global supply chains and donor interest.

- The CEO Water Mandate has established the Water Action Hub (the Hub), an enabling platform designed to address collaboration among business and other stakeholders. The mapping out of company activities on maps so that others may align projects and resources continues to drive interest from donors and local water authorities.

- The Alliance for Water Stewardship (AWS) has now launched their international (ISEAL-compliant) standard that defines a set of water stewardship criteria and indicators to guide site and catchment level interventions through to environmentally, socially and economically beneficial ends.

- CDP has created the world's first methodology for evaluating and benchmarking corporate water management efforts. Scores will provide investors, managers and others with an independent, objective, comparable, data-driven tool to assess relative company performance. It is anticipated that the development of a water scoring methodology will significantly raise the visibility of water as a strategic issue within companies and increase transparency around efforts they are making to manage water more effectively. The objective is to ensure that companies with the greatest potential to impact water resources will learn, before it's too late, how to manage water resources and reduce their impacts.

- Valuation methodologies are being pursued by a number of consultancies, NGOs and others to help take static risk scores and move them into financial risk metrics. As these techniques develop and financial arguments can be more forcefully made, companies will start to invest greater amounts into both technical and innovative strategies for the company.

There is a groundswell of opinions coming from both consumers and companies to the effect that business should do more about water. We foresee a situation in ten years where almost all companies will have a water element to their sustainability strategy and have incorporated water metrics into their risk management systems. Let's remember that the current conversations about water risk, footprint and stewardship have emerged in less than ten years. We have achieved a lot in that time and while there is still a lot of alignment and development needed before we see it fully mature, the architecture for external engagement is being rapidly assembled. Water stewardship is on the rise and is already showing promise. Its fruition provides some important opportunities for the world of business.

Important Initiatives and Tools

1. Water Risk Filter (WRF): **http://waterriskfilter.panda.org/**

2. World Wildlife Fund (WWF Water Stewardship Programme): **http://wwf.panda.org/what_we_do/how_we_work/conservation/freshwater/water_management/**

3. Alliance for Water Stewardship (AWS): **http://www.allianceforwaterstewardship.org/**

4. Water Footprint Network (WFN): **http://www.waterfootprint.org/?page=files/home**

5. CEO Water Mandate: **http://ceowatermandate.org/**

6. CDP water programme: **https://www.cdp.net/en-us/programmes/pages/cdp-water-disclosure.aspx**

7. International Council for Mining and Metals (ICMM): **http://www.icmm.com/**

8. World Resources Institute (WRI): **http://www.wri.org/our-work/project/aqueduct**

Notes

1. WEF Global Risk Report, 2014. Geneva: http://www.weforum.org/reports/global-risks-2014-report

2. McDonald's Sustainability Report: http://www.aboutmcdonalds.com/mcd/sustainability.html

3. *National Geographic*, 2010. Special edition: Water.

4. Money, A. 2012. Managing what you measure: Corporate governance, CSR and water risk. SSRN, 19 April. Available at SSRN: http://ssrn.com/abstract=2042564

5. Levitt, S.D. and Dubner, S.J. 2005. *Freakonomics: A Rogue Economist Explores the Hidden Side of Everything* (New York: Harper Collins USA), 245 pp.

6. Pegram, G., Orr, S. and Williams, C. 2009. *Investigating Shared Risk in Water: Corporate Engagement with the Public Policy Process* (Woking, Surrey, UK: WWF-UK).

7. Orr, S. and Cartwright, A. 2010. Water scarcity risks: Experience of the private sector. In L. Martinez-Cortina, A. Garrido and E. Lopez-Gunn (eds) *Re-thinking Water and Food Security* (London: CRC Press).

8. ICMM Water Stewardship Framework, 2014: https://www.icmm.com/document/7024

9. WWF, 2012. Shared risk and opportunity in water resources – Seeking a sustainable future for Lake Naivasha: http://awsassets.panda.org/downloads/navaisha_final_08_12_lr.pdf

10. Greenpeace, 2010. Dirty Laundry: http://www.greenpeace.org/international/en/publications/reports/Dirty-Laundry/

11. AWS website: **http://www.allianceforwaterstewardship.org/**

12. SABMiller and WWF, 2009. *Water Footprinting: Identifying and Addressing Water Risks in the Value Chain* (Woking, Surrey, UK: SABMiller and WWF-UK).

13. Morrison, J., Schulte, P., Orr, S., Hepworth, N., Pegram, G. and Christian-Smith, J. 2010. *Guide to Responsible Business Engagement with Water Policy* (Oakland, CA: Pacific Institute/The CEO Water Mandate, UN Global Compact).

14. Hepworth, N. and Orr, S. 2013. Corporate water stewardship: New paradigms in private sector water engagement. In B.A. Lankford, K. Bakker, M. Zeitoun and D. Conway (eds) *Water Security: Principles, Perspectives and Practices* (London: Earthscan Publications).